On a New Method of Multiplication and Shortcuts

Artem Cheprasov

ISBN-10: 1448689333

Dedicated to my mother Olga, my brother Max, and to my best friend Mocha.

A special thank you to Natasha for all of her help.

Preface

I sincerely hope that I have explained the methodologies here in an accessible manner; as with any other process, practice will give you the ability to multiply quickly on paper and in your head. The formulas and terminology may look intimidating and may seem extraneous but they're put in this book for those that need a more mechanical and rigorous way of looking at this method and the shortcuts. I may have repeated certain concepts and points more than once, not only to emphasize their importance, but also to provide more than one way of looking at the same issue in hopes that if one way of explaining it wasn't helpful, another way may be.

This book was written for anyone that may enjoy learning about unique mathematical intricacies, mental math, multiplication tricks, and others who may struggle with more traditional techniques. If you feel that you can find an alternative method to anything, by all means try, because you never know how useful it may be to someone else.

Contents

Chapter 1

Terminology

Prior to learning the new method of multiplication, please familiarize yourself with the following terminology:

-The hundreds digit of a number is represented as: "h".

-The tens digit as: "t".

-The units digits as: "u".

Since these are multiplication algorithms, two numbers are always being compared. Therefore,

-The larger number's digits have a subscript of 1 (i.e. h_1, t_1, u_1) and,

-The smaller number's digits have a subscript of 2 (i.e. h_2, t_2, u_2).

For example:

$729 * 457$

729 is represented as: $h_1 t_1 u_1$

457 is represented as: $h_2 t_2 u_2$

-If I were to multiply $u_1 * u_2$ ($9 * 7$) I would get: $t_{(u_1*u_2)}u_{(u_1*u_2)}$ which is 63. The 6 being the tens digit from the product: $t_{(u_1*u_2)}$ and the 3 being the units digit from our product: $u_{(u_1*u_2)}$.

-A capital H or T is representative of that digit being multiplied by its respective power. In this example: H_1 is 700 from $7 * 100$, H_2 is 400 from $4 * 100$, T_1 is 20 from $2 * 10$, and T_2 is 50 from $5 * 10$.

-Finally, a new symbol is introduced in order to bypass extra multiplication steps. The symbol "@" represents the attachment of one number to another.

Chapter 2

Method: $2 * 2$ Multiplication

We can dive right into the most important algorithm. Mastery of $2 * 2$ multiplication via the method presented below is necessary prior to continuing on to larger number multiplication, as well as the shortcuts in chapters 3 and 4.

The formula for $2 * 2$ multiplication is:

$$\left[\left(\left(\left(t_1 - u_1\right) * u_2\right) + \left(u_1 * t_2\right) + \left(u_1 * u_2 + t_{(u1*u2)}\right)\right) @ u_{(u1*u2)}\right] + T_1 * T_2$$

Do not be intimidated by the formula above. The five steps for this multiplication algorithm are very easy to follow and I'll show you with several examples. First, you'll see the detailed thought process behind each step; it may look messy, but it's necessary to get into the right mindset and "see" what needs to be done. Then, with each succeeding example, I'll take away the "training wheels" and you'll be able to easily picture the steps inside your own head.

Before we begin, please note that this first step, and the following two, all reference the same point in the beginning. This reference point, which will help you to remember the first three steps, is always going to be the units digit of the larger number in the pair. In the first example below, the larger number is 53. Hence, as per the instructions outlined in the beginning of this book, the subscript for both digits in 53 is 1; and the reference point is the number 3, also known as u_1.

Example 1:

$$53 * 47$$

The answer to this is as easy as: $(14 + 12 + 23)@1 = 491 + 2000 = 2491!$

Try it for yourself:

Step 1:

The first thing we need to do in the formula is to compute the following:

$$\left[\left(\underline{((t_1 - u_1) * u_2)} + \cancel{(t_2 * t_1 + (t_1 * u_2) + t(v_1 * u_2))}\right) @ \cancel{u_{(1 * u_2)}}\right] \cancel{H/\sqrt{H_2}}$$

This step is the most "complicated". As per the formula, take 3 (u_1) and subtract it from 5 (t_1), and multiply the sum by 7 (u_2).

This can be visualized as follows:

$$53 * 47$$

$$(5 - 3) * \cancel{X}7$$

$$(2) * \cancel{X}7 = 14$$

4

Step 1 is now complete.

Step 2:

$$\left[\left(14 + \underline{(u_1 * t_2)}\cancel{+(u_1*u_2+t_{(u_1*u_2)})}\right) \cancel{@ \; u_{(u_1*u_2)}}\right]\cancel{\text{H}/\text{H}/*\text{H}_2}$$

As mentioned before, start with u_1 (3), and this time don't subtract it from anything, just multiply it by t_2.

Again, this can be seen as:

$$53 * 47$$

$$\cancel{5}3 * 4 \cancel{7} = 12$$

You now have the second step's answer.

Step 3:

$$\left[\left(14 + 12 + \underline{(u_1 * u_2 + t_{(u1*u2)})}\right) \cancel{@ \; u_{(u_1*u_2)}}\right]\cancel{\text{H}/\text{H}/*\text{H}_2}$$

Step 3 and step 4 have similar calculations and hence are left next to each other for simplicity's sake. For step three, once again refer to u_1 to begin. This time, don't multiply it by t_2; instead, multiply it by u_2. Then, all you do is add the product's tens digit back to your product.

$$53 * 47$$

$$\cancel{5}3 * \cancel{4}7 = 21$$

The tens digit of your answer, 21, is 2. This is added to 21.

$$21 + 2 = 23$$

Next, add all the three final answers together:

$$(14+12+23) = 49$$

Step 4:

$$\left[\left(49\right) @ \underline{u_{(u1*u2)}}\right]\cancel{\cancel{|\cancel{T_1}\cancel{*}\cancel{T_2}}}$$

Remember the units digit from your answer from step three, before you added the tens digit back to your answer? It was 1, from 21. Now you get to use it. All you do is attach, not add, but attach it to the end of the sum of all the three previous steps:

$$49 @ 1 = 491$$

Step 5:

$$491 + T_1 * T_2$$

T_1 is simply the tens digit multiplied by 10: 50. Likewise T_2 is 40. Or you can just imagine:

$$53 * 47$$

as

$$50 * 40 = 2000 + 491 = 2491$$

In the end, to get to your answer all you had to do was:

$$53 * 47 =$$
$$(5-3) * 7 = 14 + (3 * 4) = 26 + (21+2) = 49@1 = 491 + 50 * 40 = 2491$$

With practice you'll see it in your head as,

$$53 * 47 = 14 + 12 + 23 = 49@1 + 2000 = 2491$$

and then it won't take more than a few seconds to get this result. Note the pattern of continually referring back to u_1 for the first three steps; this will help you remember what to do.

Let's move on to another example, this time simplified in its approach since you are aware of what to do.

Example 2:

$$79 * 26$$

Step 1:

Take 9, subtract it from 7, and multiply by 6.

$$(-2 * 6) = -12$$

Step 2:

Take 9 again and multiply it by 2. At this point you should be continually adding the previous answer to each succeeding one instead of waiting until the end if you have trouble remembering all three answers until step three is completed.

$$9 * 2 = 18 - 12 = 6$$

Step 3:

Once again, take 9 and multiply it by 6 and add the tens digit.

$$9 * 6 = 54 + 5 = 59 + 6 = 65$$

Step 4:

Attach 4 to 65.

$$65 @ 4 = 654$$

Step 5:

Add $70 * 20$:

$$654 + 1400 = 2054$$

Example 3:

$$43 * 90$$

This can be seen as:

$$(4 - 3) * 0 = 0 + 3 * 9 = 27 + 0 = 27@0 = 270 + 90 * 40 = 3870$$

Since you are aware of what $2 * 2$ multiplication is all about, you'll next learn a couple of shortcuts for $2 * 2$ multiplication. One shows you how to use a trick formula for any $2 * 2$ problem where both numbers end in five. The other shows you another trick formula for readily multiplying $2 * 2$ problems where both numbers are equidistant from multiples of 100 (in this case those multiples are 0 and 100).

Chapter 3

Shortcut: $2 * 2$ "Ends of Five" Formula

For $2 * 2$ multiplication, if the two numbers both end in 5, you can use a special formula to painlessly get your answer.

Example 4:

$35 * 75$

Before we get to the formula for this, there is a preliminary step you must undertake. You should always round down (borrow from) the smallest number to the nearest multiple of 10, and round up (lend to) the largest number to the nearest multiple of ten. Just remember the old adage: the rich get richer (the bigger number is rounded up), and the poor get poorer (smallest number is rounded down).

Preliminary Step:

$35 - 5 = 30$, which is represented as X.

$75 + 5 = 80$, which is represented as Y.

After you have completed this preliminary step, only use the original equation as a reference for the usual symbols; don't refer to X and Y (i.e. u_1 is still 5 from 75, not 0 from 80). Here is the actual formula:

$$X * Y + 50 * (t_1 - t_2 - 1) + 75$$

Hence,

$$35 * 75$$

$$= 30 * 80 + 50 * (7 - 3 - 1) + 75 = 2625$$

If you are more compassionate, you may round the larger number down instead, and the smaller number up:

$$X * Y - 50 * (t_1 - t_2 - 1) - 25$$
$$= 40 * 70 - 50 * (7 - 3 - 1) - 25 = 2625$$

Chapter 4

Shortcut: $2 * 2$

"Borrower's" Formula

If two numbers are equidistant from the nearest multiple of 100, another special formula may be used to quickly find the answer.

Example 5:

$$33 * 67$$

You can instantly recognize, 67 is 33 away from 100, and 33 is 33 away from 0; both are equidistant from the nearest multiple of 100. As with the previous formulas, take from the poor, and make the rich even richer.

Preliminary Step:

$33 - 3 = 30$, which is represented as X.

$67 + 3 = 70$, which is represented as Y.

After you have completed this preliminary step, only use the original equation as a reference for the usual symbols; don't refer to X and Y (i.e. u_1 is still 7 from 67, not 0 from 70).

The Borrower's Formula is:

$$X * Y + u_1 * u_2 + u_2 * (T_1 - T_2)$$

Thus, the answer is:

$$30 * 70 + 7 * 3 + 3 * (60 - 30) = 2100 + 21 + 90 = 2211$$

Chapter 5

Subscripts

As the saying goes: "All roads lead to Rome." Which path will you choose? Which is the fastest? This method of multiplication allows you to custom suit your abilities and desires to the problem at hand in any number of combinations. In the end, you'll reach the same result as another person who may have chosen a different way because they felt it was best for them.

Before showing you how to use your own strengths for a multiplication problem, please realize that mastery of the first two chapters is critical and will allow you to see the logical progression from $2 * 2$ multiplication to any length number multiplied by any other length number.

In chapter 2 I left out a couple of things so as to be consistent during the early learning process and in order to help you grasp the important steps without having to add any unnecessary confusion.

The first thing I left out is that it really doesn't matter which number has the subscript of 1 or 2. It can be either the smaller or the larger number, as long as you designate one number to have all of its digits with the subscript 1 and the other number to have all of its digits with the subscript 2. Take for instance our first example in chapter 2.

$$53 * 47$$

We designated 53 as $t_1 u_1$. Now switch it around and make 47 $t_1 u_1$ and 53 $t_2 u_2$. To solve the problem the other way you'd just use the formula presented on pg.3:

$$(4 - 7) * 3 = -9$$
$$7 * 5 = 35$$
$$7 * 3 = 21 + 2 = 23$$
$$-9 + 35 + 23 = 49@1 + 2000 = 2491$$

Having the option to label either number with the subscript of 1 or 2 allows you to have the flexibility of whether you want to deal with a negative number right away, or not deal with one at all; as shown in the first example in chapter 2 (sometimes subtraction will lead to a nice round number which is easier to deal with). In the end, you'll get the same answer, with a different approach.

Key point 1: It doesn't matter which number is designated with a subscript of 1 or 2. This give you to the option choose your own path and comfort level while multiplying. With practice, you'll quickly learn that designating one number with the subscript 1 instead of the other may help you solve a problem faster.

Chapter 6

Phases: Part 1

Now that you're comfortable with that concept, I'd like for you to stop thinking of t referring to the tens digits and u to the units digit. I'll now be referring to the t and u "phase." Since the numbers you multiply can get infinitely large, dealing with only the tens or units digits is obviously useless, and creating a different algorithm for each type of multiplication length problem is illogical and would not be a method of multiplication, but rather an algorithm limited to that particular type of problem (as per chapters 3 and 4).

Regardless, the important concepts you learned in chapter 2, of breaking a number up into two parts (then being the tens and units digits) will now help you grasp the following concept.

Key point 2: Each number, no matter how large or small, will be divided into two parts just like before. You will designate one set of digits of one number as the u-phase and the remaining set of digits of that same number as the t-phase, do the same for the other number (for the other subscript), and simply plug in the t-phase and u-phase into the algorithm from chapter

2.

This sounds harder than it really is.

Example 1:

2326 * 3829

You're multiplying a 4 digit number by a 4 digit number. You must always, no matter how long one number is and how short the other is, or if they're equal in length, have the same number of digits in the u-phase.

Key point 3: Again, u_1 and u_2 must have the same number of digits. The amount of digits in the t-phases is irrelevant and doesn't have to be the same regardless of the length of the numbers you're dealing with.

In this case, there are 3 combinations of the t and u phases I can achieve:

Combination	t_1	u_1	t_2	u_2
1	232	6	382	9
2	23	26	38	29
3	2	326	3	829

Example 2:

84938 * 234

Combination	t_1	u_1	t_2	u_2
1	84	938	0	234
2	849	38	2	34
3	8493	8	23	4

Example 3:

482008630 * 38479

Combination	t_1	u_1	t_2	u_2
1	3847	9	48200863	0
2	384	79	4820086	30
3	38	479	482008	630
4	3	8479	48200	8630
5	0	38479	4820	08630

Note how the u-phase is always the right-most set of digits you choose, and the t-phase is everything that remains over to the left. The choice of having several different ways to approach the multiplication problem in addition to designating one number with subscripts of 1 or 2 (doubling the number of combinations) will be very useful in certain scenarios to help you choose a much faster way of multiplying, as will be demonstrated very soon.

Chapter 7

Phases: Part 2

For those of you wondering about the last few examples, yes, you could technically have no digits in both subscripts of either the u or the t phase, but that would be counter-this-method as you're trying to split the problem up to multiply smaller amounts faster. Moving over both numbers in their entirety into either the t or u phase would just leave you with a typical multiplication problem.

Key point 4: The only time you would ever have only a 0 in either the u or t phase is if:

1.) The 0 is an actual digit in the number or,

2.) You may designate the smaller of the two numbers to be entirely in the u-phase. This would mean there's a 0 in the smaller number's t-phase. However, since the other number is larger, and you can only have the same amount of digits in the u-phase, that means that the larger number's t-phase has at least one digit in it. This makes sense because you can't possibly move the larger number into only its u-phase because the smaller number won't have any digits left over in t-phase, leaving you with two empty t-phases, which again, leads to a standard multiplication problem that you're trying to avoid.

Let's solidify these last two points with a different example:

Example 1:

$4827290 * 382$

Combination	t_1	u_1	t_2	u_2
1	4827	290	0	382
2	48272	90	3	82
3	482729	0	38	2

Having only one zero in one of the phases may sometimes be beneficial in multiplying larger numbers faster, and sometimes the 0 in the phase is actually part of your digit (as in combination 3 above, but unlike combination 1 where you may have chosen to move the entire number to the u-phase; but that's ok because the other number is so large that it has four digits left over in its t-phase to make sure at least one of the t-phases has one non-zero digit in it).

Chapter 8

The Full Method

After you've finished with the preceding chapters, just refer to the original method outlined in chapter 2 and proceed step by step to get your answer. Let's use a fresh example to solidify all the previous points and show you step by step that the application process of the algorithm is the same with the phases as it was with the tens and units digits of chapter 2.

$$227 * 4011$$

The most uncomplicated answer would yield the following:

$2800 + 22 + 7$, attach a 7, and add 882200 to get your result of 910497.

I'd like to, however, use some of the other combinations to illustrate a few points before getting to the answer above.

The first question is which number do I designate the subscripts of 1 or 2? Right now, you're thinking this through, but with practice you'll quickly see why one way may be better than the other. I'm already thinking about the fact that the number which I designate to have subscripts of 1 will need to have a subtraction step (step 1 in chapter 2). As always, I want to choose a path of multiplication that will offer the path of least resistance. I chose 227's digits to have subscripts of 1 and 4011's digits to have subscripts of 2. Can you guess why? It's because I'm already thinking ahead as to how I'll designate the u and t-phase. My choices are:

Combination	t_1	u_1	t_2	u_2
1	22	7	401	1
2	2	27	40	11
3	0	227	4	011

Again, with practice, this table will become second nature and you won't have to even write it out. You'll begin to notice how within a number and between the numbers you're multiplying which combination and subscript is quickest to deal with. Of course, this is all relative, as one person may find one way more desirable than another. Yet that's the whole point: you'll get the same result by taking a path of multiplication that you believe is most suitable for you.

In this case, I chose combination 2. Why? No, not because I love having negative numbers to sum up at the end of step 3 (from chapter 2), it's because for me 2-27 equals -25; who care if it's negative? For me, multiplying 25 by anything is much easier than say, multiplying 29 by something (if I chose to use combination 2 with 4011's digits with subscripts of 1, leaving me to subtract 11 from 40 for the first step of the formula).

So, as per chapter 2, I perform the following:

$$\left[\left(((t_1 - u_1) * u_2) + (u_1 * t_2) + (u_1 * u_2 + t_{(u1*u2)})\right) \,@\, u_{(u1*u2)}\right] + T_1 * T_2$$

Step 1: $(2 - 27) * 11 = -25 * 11$ [which is merely seen as $25 * 10 + 25$] $= -275$

Step 2: $27 * 40$ [again I used this method to rapidly break everything down where it's simply $27 * 4$ with a 0 attached] $= 1080$

Step 3: $27 * 11 = 297$

Now, you're probably wondering which part of step 3's answer is added back $(t_{(u1*u2)})$ and which is attached after the first 3 steps are summed $(u_{(u1*u2)})$? This is easy and I'll demonstrate right now with several examples.

Key point 5: As per the formula, you add back $t_{(u1*u2)}$ to step 3 and attach step 3's answer's $u_{(u1*u2)}$ after summing the first 3 steps together; hence: you must always attach the exact number of digits as you decided to have in u-phase. I chose combination 2, with two digits in u-phase for each number. This means that the u-phase here is also the two right-most digits (97 from 297), which are the ones I attach after summing the first 3 steps. This leaves the digit 2 in the t-phase, which is added to back to step 3. Any addition back to step 3's answer only happens if you have an answer in step 3 with more digits than the number of digits in u-phase!

Consistency is the key here, there's nothing to remember, and this point becomes second nature with practice. If you chose 2 digits for u-phase originally, and because as per the second chapter u from step 3's answer is what's attached, then only 2 digits are attached. The number in step 3's answer that's in front of those two right-most digits is added back to step 3.

For additional reinforcement of this point: had I had 5, 6, 100, etc digits for my answer in step 3, with 97 as the last 2 digits, I would still only attach 97 because that's the amount of digits in my u-phase, and I would then add the number to the left of 97 back to step 3 since that is automatically my t-phase and t is what's added back to step 3's answer.

Hypothetically:

1.) Step 3's answer with two digits in u-phase is: 38497, therefore: $38497 + 384 = 38881$. After summing up the first 3 steps, you'd attach 97 to the end of that answer.

2.) Step 3's answer with three digits in u-phase: $38497 + 38 = 38535$. Attach the 497 after summing everything up.

3.) Step 3's answer with four digits in u-phase: $38497 + 3 = 38500$. Attach 8497 to the end of the sum of the first 3 steps.

And so on.

If I have 2 digits in u-phase and my answer for step 3 was exactly 2 digits long, like 97, I wouldn't add anything to step 3's answer as there aren't any digits that remain to the left of my attachment amount which is equal to the number of digits in u-phase (hence t-phase is automatically 0). For instance:

1.) Step 3's answer with two digits in u-phase is 97. Hence, $97 + 0 = 97 +$ step 1's answer + step 2's answer = XXXX@97.

If my answer for step 3 was less than the designated number of digits for u-phase (e.g. the answer was 7), I would have to attach to the front of step 3's answer as many zeroes as necessary to have enough digits to equal the u-phase's number of digits. If I chose combination 3,

Combination	t_1	u_1	t_2	u_2
3	0	227	4	011

where there are 3 digits in u-phase, and my answer for step 3 was 7 (hypothetically), I would have to attach to the front of 7 two zeroes to make it 007, and attach those digits to the end of the sum of the first three steps. To illustrate:

1.) Step 3's answer is 7. $7 + 0 = 7$. Sum up the first three steps, they equal XXXXX. You must attach u-phase back, but it must be 3 digits long, hence it would look similar to XXXXX@007.

2.) Step 3's answer is 0, with 5 digits in u-phase. Add nothing back to step 3. Sum up the first 3 steps, and attach 5 zeroes to the end of it.

As you can tell, nothing is added back to step 3's answer above, as there aren't any superfluous digits left over for the t-phase because there aren't enough digits to fulfill the u-phase of 3 digits to begin with!

Thus, with that important digression in mind, back to step 3:

Combination	t_1	u_1	t_2	u_2
2	2	27	40	11

Step 1: $(2 - 27) * 11 = -25 * 11 = -275$

Step 2: $27 * 40 = 1080$

Step 3: $27 * 11 = 297 + 2 = 299$

$-275 + 1080 + 299 = 1104@97 = 110497 + 200 * 4000$ [which is 800000, $4 * 2$ with 5 zeros attached] $= 910497$

How did I know that I need to multiply $200 * 4000$? Well, remember in chapter 1 and 2, the symbol T stood for the tens digit multiplied by its respective power? Or, take the t-phases (2 and 40 respectively) and attach enough 0's to the end to account for the missing u-phase's digits (two zeroes in this case).

Yes, some may say that with combination 1,

Combination	t_1	u_1	t_2	u_2
1	22	7	401	1

multiplying 22-7 (15) by 1 (u_2) is even easier, and the other steps are too. Great! Just choose what is best for your comfort level. Let's see how this problem would've worked out with combination 1:

Step 1: $(22 - 7) * 1 = 15$

Step 2: $7 * 401 \ [400 * 7 + 7] = 2807$

Step 3: $7 * 1 = 7$ [nothing is added because u-phase is one digit, you'd need an answer with more digits than the number of digits in u-phase to add anything back in this step].

$15 + 7 + 2807 = 2829@7 = 28297 + 220 * 4010$ [88 with 4 zeroes attached (from $220 * 4000$) $+220 * 10 = 882200$] $= 910497$

Finally, some may have preferred combination 3, although I think this one is by far the hardest one to choose and work with, and I would've avoided it at all costs:

Combination	t_1	u_1	t_2	u_2
3	0	227	4	011

$-227 * 11 = -2497$

$227 * 4 = 908$

$227 * 11 = 2497$ [with 3 digits in u-phase, add back only the 2] $= 2499$

$-2497 + 908 + 2499 = 910@497 + 0 * 4000 = 910497$

And of course, you could take all 3 combinations above, and switch around the subscripts and perhaps realize that you like another of the 3 other ways better. In essence, it's one problem with 6 different approaches; you choose which is best for you.

So, which combination would've been the most trouble-free? For me, with subscripts of 1 for 4011, combination 1 would've yielded a very clear answer:

Combination	t_1	u_1	t_2	u_2
1	401	1	22	7

$(401 - 1) * 7 = 2800$

$1 * 22 = 22$

$1 * 7 = 7$

$2829@7 = 28297 + 220 * 4010 = 910497$

Chapter 9

Additional Practice

Here are a few more examples. I won't point out all the different combinations you can have with each example as you can now do that for yourself. I will just show the way I chose to solve it. Also, try not to look at my answer, as it's not the only way to solve the problem and it will force you to think about what the best possible combination is for a seemingly difficult problem, especially the last example I give.

Example 1:

$$522 * 852$$

$(52 - 2) * 2 = 100$

$2 * 85 = 170$

$2 * 2 = 4$

———————————————

$100 + 170 + 4 = 274@4 + 520 * 850 \ [442000] = 444744$

Example 2:

$$3601 * 5006$$

$(36 - \cancel{0}1) * \cancel{0}6 = 210$

$\cancel{0}1 * 50 = 50$

$\cancel{0}1 * \cancel{0}6 = 6$

$210 + 50 + 6 = 266@06 + 3600 * 5000 = 18026606$

Example:

$$76536 * 711$$

The numbers above may not be the "prettiest" to multiply but you can make it as easy as possible on yourself:

$(7 - 11) * 36 = -4 * 36 = -144$

$11 * 765$ [seen as $765 * 10 + 765$] $= 8415$

$11 * 36 = 396 + 3 = 399$

$-144 + 399 + 8415 = 8670@96 + 7 * 765$ [with 4 zeroes attached] $= 54417096$

Finally, to end things on a really big note: Multiply 1050 by 6 billion, 532 million, 993 thousand, 299:

$$1050 * 6532993299$$

That number is so huge it may intimidate just about anyone wishing to multiply it out, even on paper. Yet look at it closely and try to solve it on your own before looking at my answer below, it's rather easy when broken down with this method of multiplication.

Step 1: $(653299 - 3299) * 1050 = 650000 * 1050 = 650000000$ [from $650000 * 1000$ or just 65 with 7 zeroes attached] $+32500000$ [from $65 * 5$ with 5 zeroes attached] $= 682500000$

Step 2: $3284 * 0 = 0$

Step 3: $3299 * 1050 = 3299000$ [3299*1000] $+ 164950$ [3300 $* 50\text{X}- 5$, with a zero attached from the 50] $= 3463950 + 346 = 3464296$

$682500000+3464296 = 685964296$@3950$+(6532990000*0) = 6,859,642,963,950$

6 trillion, 859 billion, 642 million, 963 thousand, 950. For a number that massive you did it with relative ease. Great job!